Nature on Your Side

Aphids on your roses?
Lavender growing in the rose bed will get rid of them.

Mice in the kitchen?
A piece of mint by the mousehole will drive them away.

Gnats in the house?
Keep a castor oil plant indoors and they stay out.

Rabbits raiding the vegetable patch?
Plant some foxgloves to keep them off.

There are scores of ways to deter house and garden undesirables without damaging the natural order of things.

You don't need to buy expensive and often harmful pesticides with Nature on Your Side.

Greet Buchner Fieke Hoogvelt

Nature on Your Side

How to keep pests out of the garden and the home

Pan Books London and Sydney

Originally published 1974 by De Driehoek, Amsterdam
under the title *Milieu-vriendelijke Adviezen*
This translation first published 1977 by Prism Press
under the title *Be Nice to Nature*
This edition published 1978 by Pan Books Ltd
Cavaye Place, London SW10 9PG
original text © Greet Buchner and Fieke Hoogvelt 1974
translation © Renske van der Baan 1977
ISBN 0 330 25291 7
Printed in Great Britain by
Richard Clay (The Chaucer Press) Ltd, Bungay, Suffolk

Contents

**Part one How to keep
pests out of the garden**

1 Your garden 7
2 How to make your own compost 9
3 Good and bad combinations 13
4 The flea beetle 19
5 Dogs 20
6 Rabbits 21
7 The Cabbage root fly 22
8 Aphids 24
9 Moles 28
10 Earwigs 30
11 Rats 31
12 Earthworms 33
13 Rust 35
14 Caterpillars 36
15 The scale insect 38
16 Fungi 40
17 Slugs and snails 42
18 The mole-cricket 45
19 Whitefly 46
20 The carrot fly 48

Part two How to keep pests out of the house

1 Your house 51
2 Aphids again 52
3 The cockroach 55
4 Lice 57
5 Ants 59
6 The moth 62
7 Gnats 65
8 Mice 67
9 Woodlice 70
10 The spider 71
11 Red spider mites 72
12 The fly 74
13 The flea 78
14 The wasp 83
15 Unpleasant smells 85

PART ONE

How to Keep Pests Out of the Garden

Chapter One
Your garden

If a plant in your flowerbed or vegetable patch suddenly becomes infested with greenfly or covered in mildew, then it is obviously unhappy. Parasites will never bother plants which grow under optimal conditions and are full of life. The cells of thriving plants are far too hard and too tough for them. They are far more likely to select a puny little weakling, which is in any case struggling to survive, and therefore far easier to attack. For even though we continually fight against parasites, they do play a very positive role in nature: as garbage disposal units, which get rid of old and weak vegetation to make room for younger and stronger plants. They have only become pests as unfortunately people often prefer to grow the weaker kinds of plant because they find them unusual or beautiful or simply tasty to eat. And in such cases there will always be the risk of parasitic infection. A good example of this is the rose, which is usually very susceptible to mildew and blight; yet mildew-free roses can be had, merely by choosing varieties that are resistant to the disease.

However, even the toughest plants can suddenly start to pine and consequently become infested. There are many possible causes of this, and as a gardener you will need some quite detailed knowledge to discover the source of your particular problem. There are for instance, plants which like acid soil: azaleas, rhododendrons, heathers and the like. If they are planted in calcareous soil or given a wellmeant dose of lime trouble will follow. On the other hand, there are plants which would fancy lime every morning for breakfast, lavender for example. Some plants like the shade, others adore the sun; some can resist wind perfectly well, others catch blackfly in the slightest draught. There are clay lovers and devotees of sand. This all underlines the point that expert advice is no luxury when you are laying out a garden or buying new plants; it will prevent a load of trouble.

But even then everything can be spoilt by scattering artificial fertiliser about indiscriminately. In the first place, we have finally realized that part of the artificial substance which we use so freely will leak through into the groundwater thus polluting our surface waters. And secondly, we now know that it induces an unhealthy, hurried growth, which doesn't give the young shoots a chance to ripen properly. And it is precisely these hastily-formed new shoots which provide a nice, juicy snack for the parasites. For these reasons a careful gardener will stop using artificial fertilisers, but will replace them by an organic manure. This doesn't need to be cow dung, which, although perfectly adequate when decomposed, is often quite expensive and difficult to obtain for a town garden. A good quality compost can be used instead with the same success.

Chapter Two

How to make your

own compost

It is better and cheaper to make your own compost and this is quite easy to do. Because good and well-fertilised soil is the main weapon in the fight against all plant diseases, we will now show you how to make your own compost in a town garden.

The principle of the matter is simple. By building up the compost correctly, you allow all vegetable waste to be digested as soon as possible. The most important point is to find the right spot for your compost heap. On organic farms about one-tenth of the acreage is used for making compost. In our small town gardens this will often be too great an area, but we can quite happily make do with a much smaller one. To begin we should look for a place that is fairly unobtrusive but within easy reach; you should regularly dump all your cuttings on the compost heap. The area must be big enough for two heaps, one quietly rotting away as the other is being built-up. Thus you will always have a supply of good compost, while still being able to get rid of fresh rubbish.

We start at ground-level with the compost, directly on the garden soil. This is very important as the micro-organisms, which will have to do all the work, must be able to get into the compost from the soil, and the dozens of earthworms that will develop in the compost must be able to get back into the garden. First rake the soil well and put down the initial layer of refuse. This can be anything; fallen leaves, grass mowings, twigs, or old plants. It doesn't really matter what you use as long as nothing is over four inches long. Flower stems, for instance, should be chopped up before being placed on the heap. Potato peelings should first be treated against sprouting, and grass cuttings mixed with some coarser material, otherwise they tend to stick together forming an airtight mat. Tough leathery leaves, such as chestnut or maple, must be added piecemeal, mixed with other rubbish. Twigs and woody stems rot very easily as long as they are cut fine and short.

When the pile is about eight inches high you should sprinkle a thin coating of lime on it, then add a layer of bone meal. (You can get this from garden shops). It can also be mixed with feather meal or blood meal and should be applied at the rate of eight ounces to the square yard. Bonemeal is very important as it provides the compost with the right amount of nitrogen. Cover all this with about half an inch of earth, preferably real clay if you can get some, or even ordinary garden soil with modelling clay mixed in. If you have been making compost for a while you can also take some already-decomposed rubbish from the second heap. This layer should be pressed down lightly with a rake and then the whole process is repeated: eight inches of chopped garden refuse or kitchen waste, a sprinkling of lime, eight ounces of bone-meal per square yard, and half an inch of clay or earth.

The sides of the compost heap should slope slightly to allow rainwater to drain off. Once the heap is three to five feet high it must be coated in a layer of grass cuttings to keep out daylight and wind. Now it is up to the micro-organisms, and they know their job. They'll leave a fragrant loose soil that will be ideal for your garden.

In summer the decomposing process will take about three months depending on the weather, but in the winter months it will take a little longer. This is not really a problem as there is little to do in the garden at that time of the year.

Home-made compost is the best fertiliser you can use for your garden since it is rich in all sorts of spores and micro-

10

organisms which will activate the soil around the plant. Don't dig the compost over, just spread it around the plants in a thin layer and cover with some earth. Used around fruit trees, the layer should be a bit thicker. If you treat each fruit tree to a four inch layer in autumn, the soil around it will recover in a couple of years from the ill effects of any previous chemical treatments. It would be beyond the scope of this book, which is concerned with natural ways of keeping down plant diseases, to enlarge upon the difference between the usage of artificial fertiliser and compost. However experiments carried out by Alwin Seifert on organic farms have clearly shown that well manured crops are far more resistant to pests than crops that have been artificially fertilised. Along with a healthy soil, unspoilt by chemicals, combinations of plants often play an important part too. Some plants go well together, while others do not mix at all. The reasons for this are not completely understood, but a lot of experiments have been carried out in organic gardening which give some useful pointers to the right combinations. Such combinations also prevent infection by insects and fungi; so again, we will no longer have the problem of fighting these without polluting the environment.

Sowing Dates

The sowing date influences the correct development of the plant to a great extent. Some plants grow better in the Spring, others in the Autumn; some should not be sown at certain times of the year because of the high risk of infection, carrots for instance. Others, like spinach, should not be sown in the middle of Summer because they will go to seed very quickly. Some people claim that the position of the "heavenly bodies" is of relevance here. There is a special astrological calendar produced annually indicating the best sowing dates for a great many crops.

Chapter Three
Good and bad
combinations

Careful gardeners who put their hearts into the job and are observant about their work have discovered over the years that some plants have a beneficial effect on each other, others a detrimental effect.

A great many experiments in organic farming have been done with good and bad combinations, but it will take years of thorough research to collect absolutely conclusive data. It is not always known why one combination is beneficial, while others are detrimental. However some reasons are obvious: plants with deep roots make the soil penetrable for plants with shallow roots, so they form a good combination. Similarly plants which like slightly shady ground feel very comfortable amongst taller plants, e.g. strawberries amongst runner beans. Another good type of combination is where one plant keeps the parasites away from its neighbour. Hyssop for instance does this, when planted with cabbage, and marigolds with roses. Obviously plants like legumes, which can fix nitrogen from the air, will benefit their neighbours by supplying them with this important plant food. In this way, peas and beans form a very good combination with, for instance, potatoes.

There are also plants which have a distinctly symbiotic relationship, e.g. the horseradish and the potato.

Listed below are a number of good combinations tried and tested by "biodynamic" gardeners.

Asparagus:	— with tomatoes and parsley
Beans:	— with carrots or cauliflower
	— with potatoes, particularly early varieties
	— runner beans with celery
Beetroot:	— does well when grown near onions, runner beans or swedes
Broad-beans:	— with dill, which prevents blackfly
Cabbages:	— with hyssop, late potatoes, dill, camomile, golden rod and sage. (Particularly sage)
Carrots:	— do well when planted close to lettuce
	— with spring onions between the rows will not be troubled by carrot fly.
	— with peas or dill
Celery:	— with leeks is a good combination
	— between cabbage, keeps the cabbage free from pests
Cress:	— with radishes becomes more fragrant, and the radish will have a milder taste. In the case of a flea beetle epidemic the radishes will not become infested because the flea beetle prefers cress
Dill:	— does well between carrots, lettuce and onions but do not let it flower, or the growth of the roots will be stunted
Lettuce:	— with rows of chervil, gives good dense hearts and stays free from aphids
Mint:	— *Mentha piperita* produces twice as much essential oil when grown between rows of nettles
Onions & Leeks:	— form a good combination with carrots, beetroot and tomatoes
	— camomile between these crops keeps the onion fly away
	— parsley makes the onion more fragrant
Potatoes:	— with African marigold (*Tagetes patula*). This prevents the notorious potato eel-

14

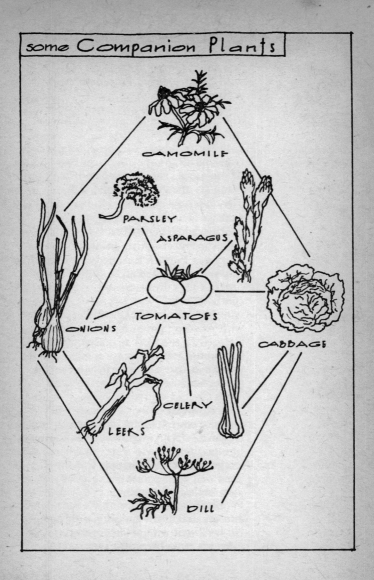

some Companion Plants

CAMOMILE

PARSLEY

ASPARAGUS

ONIONS

TOMATOES

CABBAGE

CELERY

LEEKS

DILL

worm
- with some horse radish at the corner of a bed keeps the potato healthy
- with hemp keeps *Phytophtera* away
- with peas. It is best to have two rows of peas and two or three rows of potatoes. The peas provide the nitrogen for the soil
- with beans, which, like peas, fix nitrogen from the air. Every other row plant a row of tomatoes. The root-saps of the potato stimulate the growth of the tomato.
- early potatoes like the company of the cabbage after being ridged up. The cabbage too benefits from this "coupling".
- with sunflowers. A traditional combination from Eastern Europe, where sunflowers are planted around the edge of potato fields. Both species profit from the combination.

Roses:
- with lavender prevents aphid attack. with African marigold, prevents rose fatigue of the soil. The marigolds should be planted among the roses every three years, but do not necessarily have to flower because it is the roots that do the work.
- with chives or onions the roses will become more fragrant and there is less chance of aphid infection. This combination is often used in Bulgaria where roses are grown for the perfume industry.

Spinach:
- does well after broad-beans as a catch-crop before cabbage

Strawberries:
- in combination with runner-beans, lettuce and spinach, but especially with borage. Do very well with some pyrethrum. This prevents strawberry diseases.

	— with pine needles spread between the rows, preferably the *Picea* or *Pinus* varieties.
Tomatoes:	— are a strange crop. They can stay on the same soil for years and can be fed with compost made from tomatoes. Nettles and parsley stimulate the growth of the tomatoes.
	— African marigolds prevent tomato eelworms.
Tulips:	— may never be grown on the same soil for two years in a row unless you grow African marigolds on the same soil in Summer. These keep the tulip eelworms away.

Bad Combinations

It is just as important to avoid certain bad combinations. If you put the wrong crops together you are more likely to have trouble with pests. Combinations you should avoid at all costs:

Beans:	— with onions do not form a good combination
	— gladioli are sworn enemies. Even at a distance of ten yards gladioli have a bad effect on beans
	— runner beans and fennel do not get on well together: they stunt each other's growth
Cabbage:	— following radishes or vice-versa will seldom be a success
Cauliflower:	— will not produce a good crop after spinach
Lettuce:	— in the neighbourhood of parsley will soon become infected with aphids. Also the parsley will not do well
Peas:	— peas, like beans, don't like onions or spring onions
Potatoes:	— are not successful when planted near birch

17

trees

Roses: — grown near gooseberries are bad for both
— do not like box *(Buxus sempervirens)* because its roots are obstructive

Bad Company

PEAS

LETTUCE

ONIONS

CABBAGE

RUNNER BEANS

PARSLEY

FENNEL

STRAWBERRIES

TOMATOES

Strawberries: — and cabbage do not flourish near each other
Tomatoes: — will not grow with fennel, or kohlrabi

Chapter Four

The flea beetle

The flea beetle, of the *Halticini* family, is not a flea at all, but just happens to be quite a jumper. Both beetle and larvae feed on vegetative material, and this would be no problem but for the fact that they often regard our own carefully grown vegetables as a very suitable food supply. In dry periods, when the soil is very hard and crusty, the chances of a flea beetle outbreak are high. These insects are very fond of radishes, but can be side-tracked for a while by sowing some cress between the rows of this crop, which the flea beetle likes even better.

Cold Water

Using cress is only a temporary measure and it's better to water the beds regularly with cold water. Since neither the adult beetles nor the pupae in the soil like it, this usually keeps them under control.

Board with Tar or Treacle

Another way to deal with the pest in smaller gardens is to catch them with a sticky board. The underside of the board should be covered with tar, or treacle or golden syrup, if you can't get hold of tar.

With the sticky side down you wave the board over the infested beds. This will make the fleas nervous and they will start jumping and get stuck to the board.

Chapter Five

Dogs

The best way to deal with dogs in the garden is by building a good fence around it, but where the size or shape of your garden makes this impractical the first step is to find out where the dogs usually enter. By blocking these entrances with plants that the unwanted visitors would rather avoid, you will soon be rid of them.

Common Rue

One of these plants is a perennial that dogs dislike intensely. For some unknown reason they will keep well away from common rue, and you will also have a very elegant plant.

Berberis

A thorny hedge such as berberis is also quite effective in discouraging dogs, but these shrubs should be planted very close together as the intruders may try to squeeze through any small opening.

Chapter Six

Rabbits

If you live in the country you may find your garden is popular
with rabbits. In this case a good fence is no luxury, but a row
of plants poisonous to rabbits can also make a good line of
defence.

Digitalis

Your best line of defence will be *digitalis purpurea,* the fox-
glove, a strong biennial plant. It is self-seeding, so once planted
it will maintain itself over the years, as long as its demand for
moisture and shadow is met.

Onions

You can also protect part of your garden by laying a little
hedge of two rows of onions around it. Left to develop good
thick leaves these will deter rabbits, which will not have the
stomach to eat through it.

A Good Choice of Plants

In an ornamental garden you can avoid trouble by using only
those plants that rabbits won't touch.
   For example they love silver firs, but turn their noses up at
ordinary conifers; yew trees are safe too. Bulbs, however, are

out of the question if there are any rabbits around; they would be devastated in no time because they are so delicious. Fortunately primulas are not very popular so you can use these for some colour in early spring.

Chapter Seven

The Cabbage root fly

How to Collar the Cabbage Root Fly

The cabbage root fly is a very unpleasant character and its maggots can do a lot of harm to our cabbage plants. Fortunately there is a simple remedy which will restrict its activities: the cabbage collar.

In the past, before chemical sprays were invented, cabbage collars could be bought in garden shops. Now you have to make them yourself. Cut a circle four inches in diameter out of tar paper. Make a hole in the centre large enough to give the plant room to grow. Next, cut a slit so you can slip the collar around the plant. The soil around each one should be built up a little, and the collar fastened with a peg. The cabbage collar will prevent the cabbage root fly laying her eggs on the vulnerable root sheath.

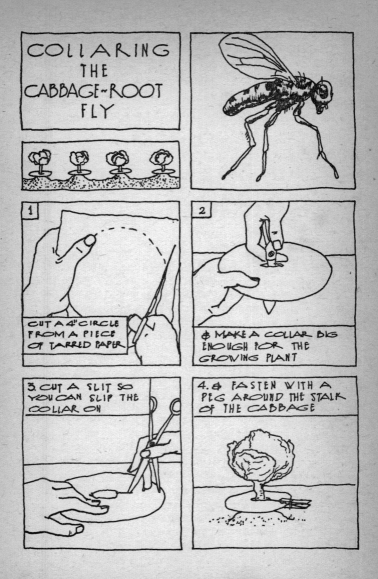

COLLARING THE CABBAGE~ROOT FLY

1. CUT A 4" CIRCLE FROM A PIECE OF TARRED PAPER

2. & MAKE A COLLAR BIG ENOUGH FOR THE GROWING PLANT

3. CUT A SLIT SO YOU CAN SLIP THE COLLAR ON

4. & FASTEN WITH A PEG AROUND THE STALK OF THE CABBAGE

Chapter Eight

Aphids

Aphids can fly for hours on end

Aphids or plant lice are very peculiar insects. In early spring the first aphid eggs that have survived the winter hatch out. These pioneer ones are always female, and can reproduce themselves for generations without any help from male aphids. Funnily enough, the males don't appear to be entirely superfluous, as in the autumn male aphids are also produced. These make sure, in their own way, that the next generation of eggs will be fertilised. Only fertilised eggs seem able to survive the winter, hence the late appearance of male eggs.

Females as well as males, develop into flyers when a plant becomes too overcrowded. Suddenly all new aphids will be equipped with a pair of wings and these winged animals are capable of flying for hours on end, in search of new food supplies. They will prefer a plant that is not really thriving, or that has just had an extra big dose of nitrogenous fertiliser. Avoidance of artificial fertiliser reduces the risk of infection by aphids. However under certain weather conditions and because we mainly use monocultures, they form a constant threat to our crops.

Nettle Manure

As a remedy against aphids we can use a nettle manure. To

make this, keep a bunch of nettle plants in a bucket of water for five days. The liquid can be diluted up to five times and is sprayed on the infected plants. If you have only a few plants that need treatment, you can use a dish-mop to sprinkle the infected places with manure which, by the way, becomes more effective the longer you leave it.

Nettle Extract

This aphid repellent will be ready within twenty-four hours. Gather two pounds of nettles and cut them up, leave them in two pints of water for twenty-four hours, then sieve. The liquid has to be applied very generously. Another application about five days later will get rid of any persistent attackers.

Topping Broad Beans

Broad beans suffer badly from one type of aphid, the black-fly, but this need not be a problem: when beans are almost fully grown you just pinch out the young tops. This will make a great difference as it is the succulent tops that the black-fly likes best of all. After topping it is best to spray with the stronger five day nettle manure as a preventative measure. As this manure contains a lot of silicic acid it also stimulates the growth of the leaves. It makes them tougher and thus less susceptible to black-fly; moreover the manure will kill the maggots of other insects too.

Rhubarb Soap

An excellent way to free your plants from aphids is by using a mixture of rhubarb extract and soapy water. For this remedy you need two pounds of rhubarb leaves and two pints of water. Boil the finely shredded leaves in the water for about half an hour, then sieve off the liquid. In the meantime make a rich lather from two ounces of soft soap and one pint of water. Mix this with the rhubarb extract. Apply it to the infected plants with a spray can or your faithful dish-mop and the aphids won't have a chance.

Ladybirds

A loyal ally in your anti-aphid campaign is the ladybird. The larvae as well as the adult ladybirds are very fond of aphids. In the USA you can order these bright little beetles by the kilo to put on your infested fields. We haven't yet reached this stage in England, but you can catch them yourself quite easily. Try to gather as many as possible and put them on your infected plants. The ladybird will be glad of this proffered meal, and the plant of the loss of its pests.

Hyssop, Sage and Thyme

These strongly scented herbs also act as aphid repellents. If you plant them spread out through your border, the chances of infection will be greatly reduced.

Lavender and Onions

Roses are especially susceptible to aphids, and with these vulnerable plants prevention is better than cure. This can be achieved quite easily by planting lavender or ordinary onions near the roses. Of course the lavender is more attractive to look at than the onions but the latter have the advantage of intensifying the smell of the roses, however odd that may sound.

Birch Trees

Recommendation by experienced farmers:
  a birch tree near a field of beans will draw off aphids,
  which prefer the leaves of the tree to those of the bean.

Dill

Dill between broad beans keeps the aphids at a safe distance and it can also be used in the kitchen, when either fresh or dried. So this remedy is definitely worth a try.

Artemisia Absinthium

If you have room in your garden for a rather stout plant by the name of *artemisia absinthium*, commonly known as wormwood, you will always have the makings of an excellent natural pesticide. You have to pick some fresh leaves from this plant before nine o'clock in the morning, and before the plant has reached the flowering stage. Add water to cover, and bring to the boil. Let the mixture cool, sieve off the liquid and dilute in a proportion of one to four. It is of great importance to stir the liquid for ten minutes. Firstly stir clockwise, creating a whirlpool effect, then anti-clockwise and so on. A well-stirred spray will be far more effective than a haphazard mixture, so stir conscientiously and don't lose patience.

Savory

This is another plant aphids cannot stand. An occasional row of savory between the beans will mean less aphids and another aromatic herb to add to some of your dishes.

Chervil

Lettuce, too, falls victim to aphids, but can easily be saved by some neighbouring rows of chervil. They will discourage the aphids. Stick to chervil though, and don't try another umbellifer such as parsley, because that has exactly the opposite effect. Nobody knows why, but it's a fact!

Chapter Nine

Moles

Moles are real insect hunters

However unlikely this may sound to the anti-mole brigade,
the mole is one of the most useful animals in agriculture.

Moles, or *Talpidae*, are real insect hunters. They consume
those larvae, maggots and caterpillars that live in and just
above the soil, and by doing so the mole helps to maintain
the balance of nature. This is of the greatest importance, even
in that little bit of nature which we call our garden. We should
never try to kill off as many animals as we can; the real aim
should always be to interfere with nature as little as possible.
If we murdered all moles, we would be declaring an open
season for pests, which we would then have to kill off by
other means, often producing unfortunate side effects.

It does not follow however, that we should just let the
little furry animals go ahead, digging their tunnels as they
please. (They can reach a speed of eight inches per minute!)
A mole can do a great deal of damage particularly in seedbeds or other beds with small, delicate plants. On the
other hand, a mole that happens to dig a tunnel in a lawn or
border doesn't do very much harm. Simply trample the mole-
hills down and all traces of underground activity will have
disappeared within two or three weeks.

What do you do if they have chosen your nursery beds?
Then you **will** have to interfere, even though it is sad for the

little animals, who have a maximum lifespan of only three years.

Carbide

Rather than using cruel traps you can place pieces of carbide in the tunnels. Because of the moisture in the soil carbide gas will develop, and this will drive off the entire mole family to other places where they can do their useful work in peace.

Camphor Balls

With the help of real camphor balls (not those surrogate chemical ones) you can keep the moles away from certain parts of your garden. You do this by constructing a "fence" of camphor balls. Every ten inches you bury a ball two inches in the ground. No mole will brave such a horrible smell.

Euphorbia Lactea

A row of milk spurge plants *(Euphorbia lactea)* will also keep the moles out of your garden, and thus protect any plants behind it.

Chapter Ten

Earwigs

The earwig, *Forficula auricularia,* takes better care of its young than most other insects. The female guards her eggs until they hatch and uses her strong pincers to defend not only the eggs but also the young. Earwigs, like woodlice, are real scavengers. They live off semi-decomposed remains and insects, and do a very useful job in this way. So generally the earwig is not harmful and can be left in peace.

Only occasionally do they do some damage — for instance to the buds of dahlias or to soft fruits. If this damage should become significant it is best to catch the earwigs.

Pots with Woodshavings

There is a very simple way to do this as earwigs are rather stupid. At dawn, they hide away from the light in the first place they can find. An upturned flower pot filled with woodshavings is an ideal place for them. We can easily catch them by putting out some of these pots around the flowers or fruit bushes at night. The next morning merely burn the wood shavings or better still empty the pots in a place where the earwigs can do no harm.

Sunflower Stems or Paper Tubes

A very elegant way of catching earwigs is to put hollow stems of sunflowers, or paper rolled to form tubes, in places where the earwigs are a threat to our plants. But again it must be stressed that this is seldom the case.

Chapter Eleven

Rats

The rat is one of the most dreaded rodents in the country, not only because of the damage he does but because he is so cunning. Throughout the ages he has been hunted down, and more and more ingenious methods have been invented to kill him, consequently only the cleverest and strongest animals have managed to survive and multiply. If you have a real plague of rats you will have to notify the local authorities, who will take the necessary action. If you have only the occasional rat in your garden you can cope with it yourself. You will have to chase him and his undoubtedly numerous relatives away. You can do this with the aid of plants that rats hate.

Valerian

Valerian, or *Valeriana officinalis*, is one of these plants. Strangely enough, cats are attracted by valerian, and this may help to drive the rats away. If you put a few of these strongly smelling plants in places where you think the rats enter there is a good chance that they will stay away.

Euphorbia and Fritillaria

Euphorbia (or spurge) and *Fritillaria imperialis* aren't rated very high by the rat family either. Planting a few of these in strategic places will do the trick.

Onions

The vole *Arvicola terrestris*, a rodent found mainly in the vicinity of water, is very fond of the roots of young trees and other plants. Its gluttony can do serious damage, therefore it is advisable to surround threatened trees and plants with a row of ordinary onions. This will soon make the vole lose its appetite.

Rags Covered in Tar

If you know where the rat has its nest, the simplest solution is to block off the entrances with rags covered in tar. Rats have such an aversion to tar that they will soon go away.

Guinea Fowl

A cheerful but rather noisy way to keep rats at a distance is to buy a guinea fowl. These graceful birds make such a din that they will even frighten rats. A neat solution, as long as you don't mind the noise yourself.

Alum

As a last resort you can of course use a poison. A batter of flour and water with a large dose of finely ground alum mixed in will be readily eaten by rodents, and will quickly kill them.

Chapter Twelve

Earthworms

Earthworms are Indispensable Friends

The earthworm, one of the *Oligochaeta* family, is the most useful creature in our garden. It aerates the soil by its constant digging and grubbing, thus allowing plant roots to breath more easily. It has a remarkable intestinal system, which never stops working. Not only does it transform vegetable remains into inorganic material, it also digests large quantities of earth.

It was Darwin who discovered, after an intensive study of the earthworm, that the earthworm population in one hectare (2.5 acres) of well maintained land, will digest twenty tonnes of earth per year; an invaluable contribution to the improvement of the soil. And although they live on vegetable matter, they don't do any real damage to our crops. They can be harmful to young seedlings but this damage is only to certain parts of the garden and of a temporary nature, so drastic measures are unnecessary. It is enough to 'divert' their activities. In order to do this, entice them to the surface so you can catch them and transfer them to another part of your garden — or better still to your compost heap where they will make themselves very useful. Never destroy earthworms; it is thoughtless and wasteful to do so.

You can solve the problem by transferring your seedbeds to another part of the garden, but there are also quite a few good household methods of removing them locally.

Wood Ash

A very simple method is to use wood ash. Before sowing, scatter some wood ash on your seedbeds. Because the earthworms do not like this, they will come to the surface, where they are easily caught. This technique is particularly effective in wet weather.

Chestnut Spray

During the conker season you have another good method at your disposal. Take ten horse-chestnuts and boil them for an hour in two pints of water. After they have cooled down, take the chestnuts out and use the water to spray on the seedbeds. As with ash, the worms will soon come to the surface.

Nicotine

A less friendly way of dealing with them is by using a nicotine-extract. This is harmful to the worms, so use it only as a last resort, and remove the worms from the sprayed area as quickly as possible. To make the extract you put a small handful of tobacco in a bucket of water and leave it for an hour.

Valeriana Officinalis

If you want to attract earthworms to improve the soil structure, your best bet is to water the soil once a month with an extract of Valerian. Pick the leaves in the morning, cover them with water and bring to the boil. When it has cooled down, dilute the extract with water in the proportion of one to four. Then stir for ten minutes, first clockwise then anti-clockwise. It is important that the extract is properly mixed.

Never use it near pot plants, as they may become damaged by a surfeit of earthworms around their roots.

Chapter Thirteen

Rust

Rust is the name of a brown fungus, which attacks many crops. As with all other plant diseases prevention is better than cure, and prevention means growing your crops on a good healthy soil, in the right combinations using a proper rotation pattern. .

However, these factors will not always be enough to prevent rust on your plants, because it is a very peculiar type of fungus. It likes to move around. Most rust fungi hibernate on completely different plants from the ones on which they feed in summer. For instance, the type of fungus which can do severe damage to cereal crops, likes to spend the winter on the Barberry bush (*Berberis vulgaris*), while the fungus that causes the familiar pear-rust, moves to the Juniper in winter. And the rust that attacks French beans survives on various conifers. So if we want to prevent the rust, we will have to keep these winter homes well separated from our crops.

Beans

Farmers know about this. You will never see an experienced farmer plant his beans anywhere near conifers, because that would greatly increase the chance of rust damage.

Pears

Lovers of pears will have to avoid juniper bushes in their garden. If you grow both you'll have to watch very carefully for any signs of rust on the pears.

Chapter Fourteen
Caterpillars

The caterpillar is like the ugly duckling, a clumsy, greedy youth, which will eventually develop into a very elegant creature. To become a butterfly, the tiny caterpillar which leaves the egg has to eat itself into a fully grown one in a very short time; then it must spin itself into a cocoon to undergo the last stage of the transformation. The butterfly is completely harmless to our plants, and is often a colourful addition to the beauty of the garden. But it will soon start to lay eggs, and when these hatch the problems of the caterpillars' rapacious appetite start all over again. Therefore we have to do something about the caterpillars, if we want to safeguard our crops from their hungry attack.

Cabbage Caterpillars

If you see any cabbage whites frolicking about in your garden be on your guard, because it is highly likely they will have laid their eggs on the underside of the leaves of your cabbages. In a small garden it does not take long to check the cabbage leaves regularly for eggs, and to crush those easily visible clusters between your fingernails.

Derris and Pyrethrum Powder

An efficient remedy against caterpillars is a mixture of two vegetable powders: derris powder, made from the roots of an exotic labiate, and pyrethrum powder, which comes from a well-known composite flower.

Be careful, though, when buying this powder, because you might easily be sold a chemical which could upset the natural balance in your garden and kill useful insect-eating birds.

Bacterial Preparations

A bacterial agent is now also available against caterpillars. It is very selective, and kills them without harming people, animals, or other insects. But again, make sure you get the right concoction.

Grass

This is one of the most amazing tips sent to me. A gardener, who had been plagued for years by caterpillars on his cabbages, lost his temper one morning, and dumped his grass mowings all over them. Miraculously the caterpillars disappeared within a few days. Encouraged by this, he kept scattering fresh grass over his cabbage plants each time he mowed the lawn, and in following years he was never troubled by caterpillars again.

Coincidence? After so many years of caterpillars, followed by just as many without, but with grass, it seems hardly likely.

Anyway such a harmless remedy is always worth a try. But don't use too much grass, as it will start rotting. A thin covering once a week will be enough.

Chapter Fifteen

The scale insect

The Scale Insect Hides Her Eggs Under Her Own Scales

The scale insects, of the *Coccidae* family, are not easily recognised. Beginners especially may not notice them. The female is a brownish domed spot oh the leaves and stems, and the male is even harder to recognise. He looks more like a small gnat, flitting about the plant.

Keep a watchful eye open for them, because they may soon prove to be a real plague, given half a chance.

The female scale insect may look quite inoffensive, like immobile brown domes, but their eggs are safely hidden under their waxy scales; and when mother dies this scale is absolutely packed with eggs, from which the larvae hatch. In this way they multiply rapidly, producing in a short space of time a multitude of brown scales, which feed on and damage our plants.

Soap and Methylated Spirits

Fortunately it is very simple to control scale insects, especially if only a few plants are infested. In this case simply remove all scales from the leaves with your fingernail. Repeat this treatment after a week.

If too many plants are infested for you to check all the

leaves individually, the following mixture will help:
two pints of water, one teaspoon of soft soap, and one
tablespoon of methylated spirits. Spray your plants with
this mixture or brush it over the leaves. Rinse the leaves
the following day and repeat the whole treatment a week
later.

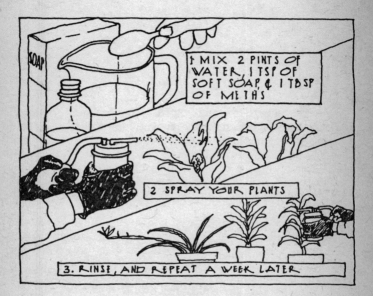

Chapter Sixteen

Fungi

The name 'fungus' is used for such a large group of primitive
plants, that you would hardly recognise some of them as being
related. For example the toadstool is a fungus, but so is the
mildew which covers the leaves of your roses with a thin white
film. Some fungi are completely harmless to plants and animals
(one of these, *Penicillium*, even provides us with one of our
best known medicines), but others may cause severe damage
to plants.

Among the latter are the well-known rust fungi, which we
have mentioned before. Equally harmful is the fungus that
causes blight in potatoes. This particular organism, which was
responsible for the great Irish famine of the last century,
can only be prevented by crop rotation and good healthy soil.

Lastly, there are the wide range of different types of
mildew which can attack your plants. Suddenly you will find
that the tops of some plants, roses for instance, are covered
in a white fluffy layer; by this time, however, it is already too
late. It is virtually impossible to fight fungus infections. All
you can do is try to prevent them in the first place by proper
manuring, that is without chemicals, and in the second by
choosing varieties that are most resistant to mildew.

Onion Extract

Finally, you can use an onion extract, as a preventative

measure against mildew.

Before you start, cut out the tops of the mildew infested plants, as these are the most vulnerable parts, the more so if you have used some artificial fertiliser! (But you will be wary of that by now).

Once the tops have been cut out and burnt (the safest way to prevent any further infection), spray the suffering plants with the onion extract.

You make this by putting a large bunch of onion leaves (don't use the bulb) in a bucket of water for four days. Then sieve this pungent liquid, and apply it to the plants with a spray can or dishmop. Repeat the treatment once a fortnight, starting as soon as the plants begin to form new shoots until the middle of summer. It is the best way of preventing mildew attack.

Sulphur

An old-fashioned preventative method is to spread sulphur powder over the leaves of those plants which are likely to suffer from mildew. The drawback of this method is that the yellow powder all over the leaves looks nearly as bad as the mildew itself.

Equisetum

An excellent way of preventing mildew is by spraying regularly with a tea made from dried horsetails or *Equisetum*. You can either collect and dry these yourself, or order them dried. For one treatment you need one ounce of dried horsetails in two pints of water. Boil this mixture for at least half an hour. If possible leave it to draw and strengthen for another two days before sieving, then dilute in a proportion of one to four. Also, give it a good stir before using. Ten minutes, first clockwise then anticlockwise, will make all the difference.

Chapter Seventeen

Slugs and snails

Slugs and snails, the *Gastropoda*, may be very interesting animals, but we can do without them in the garden because they are really greedy. The presence of slugs is usually revealed by the gaps that suddenly appear in leaves. A shining, silvery trail will confirm your suspicion that you are dealing with slugs.

Again, with slugs as with other problem animals: don't kill them if you can avoid it.

If you want to safeguard your plants from their ravages, you can in many cases keep them off by surrounding the threatened plants with others to which slugs have an aversion.

Defence Plants

These can often prevent a lot of disappointment. One example is common Sage *Salvia officinalis*. But be sure to use the right species.

Salvia splendens or scarlet sage is one of the slug's favourite dishes, and would have exactly the opposite effect.

Hyssop, *Hyssopus officinalis*, and all varieties of thyme will also form a good defence line.

Extract made from Fir Seeds

In organic horticulture, slugs are often controlled by an extract made from fir seeds. It is applied in a strongly diluted form (3:1000), which keeps the slugs away very effectively without killing them.

Beer

A much less friendly method to catch the slugs is by using beer. They are extremely fond of beer and will willingly drown themselves in it. Some people sink a half full jar of beer in the ground close to the threatened plants; others put down a full soup plate.

In either case the slugs will come out for a drink at night and drown in style.

Rhubarb Leaves

If you have a large garden, you are likely to have rhubarb in it somewhere. Pick a few of the leaves in the evening and place them upside down among the damaged plants. At dawn the slugs will seek shelter under these, and all you will have to do the following morning is bring them, leaf and all, to the compost heap.

Slug Poison

A mixture of ten parts wheat-bran and one part finely ground *Meta* block, (available from hardware shops), placed in saucers on the slugs' route will prove fatal to them. Most slug pellets, bought readymade, contain the same mixture. But make sure, if you buy any of these that there are no other more harmful poisons among the ingredients listed on the label.

Never scatter the pellets over the plants, just place some around on the ground. And wash your hands thoroughly afterwards.

Wood-shavings or Straw

Moistened wood-shavings or a few handfuls of damp straw can be put around the plants you want to protect. The poor unsuspecting slugs will hide away in here at sunrise and can be easily caught and removed.

Pebbles, Crushed Egg Shells or Chalk

Precious plants which are often loved by slugs — and also seedbeds — can be protected by surrounding them with a thin layer of pebbles, crushed eggshells or chalk. Slugs are unable to cross such a barrier and so it is an ideal defence.

Chapter Eighteen
The mole-cricket

The mole-cricket, or *Gryllotalpa*, is a cousin of the cricket and is really a very useful animal, although being about two inches long it may appear rather frightening. It is a typical burrower and lives, like the mole, under the ground for most of the time. There it does a very useful job in catching and devouring all insects, larvae and cocoons that it meets on its way.

Therefore it is a pity that he damages our seedbeds from time to time. When that happens you will have to catch him. This is quite easy, as the mole-cricket is rather slow-witted.

Flower Pots

All you have to do is to sink some medium-sized flowerpots over the beds keeping the brims level with the ground. Your mole-cricket will not suspect anything until he suddenly topples over the edge into the pot.

That is where you will find him the following morning, unless the blackbirds have got there first. They soon notice what is going on, and will have helped themselves to a tasty breakfast, before you have even woke up!

Chapter Nineteen

Whitefly

The Whitefly, or *Trialeurodes vaporariorum*, belongs to the *Aleurodidae*. The clear white appearance of this tiny animal is caused by a fine layer of white powder which it secretes all over its body.

The flies are actually less harmful than the immobile larvae, which live on indoor plants and feed on their sap. A badly infested plant will be having a rough time and suffer accordingly.

There is some dexterity required in the fight against this little fly, (which in fact is a bug and not a fly).

Nettle Manure

Begin by making some more nettle manure. Put a bunch of nettles in a bucket, cover and leave for four days, then filter. Fill a spraycan with this manure, and put the infested plant, (often a fuchsia, which is a real delicacy for whiteflies), on a plastic bag, which has been rolled back in such a way that you can quickly draw it up to enclose the plant.

During these preliminaries it may seem that you have chased the little flies off the plant but in fact they will be back before you are ready to start the treatment itself.

Now try to pull the bag up quickly, so the flies don't get a chance to escape. Then spray a fair amount of the nettle extract through a little opening in the bag over the leaves; ensuring that the underside of the leaves are reached as well, because that is where the larvae will be.

When you have accomplished this, tie up the bag and leave it for five days. In most cases you will then have got rid of your whitefly pest, especially if you also move the plant, so that any flies which did manage to escape cannot find their way back to the host.

Sometimes it may be necessary to repeat the treatment a few days later.

1. CUT 2 LBS OF NETTLES
2. CHOP THEM UP
3. LEAVE FOR 24 hrs IN 2 pts WATER
4. SEIVE

Nasturtiums

Whitefly on tomatoes can be controlled very easily. Just plant some Nasturtiums (*Tropaeolum*) or African Marigolds, (*Tagetes erecta*), preferably the *patula* variety, between them. The strong scent of these plants penetrates the soil and is taken up by the tomato, which then loses its attraction for the fastidious whitefly.

Chapter Twenty

The carrot fly

The little carrot fly is dreaded as it can cause havoc in our carrot beds. Once again, prevention is better than cure — and in this case prevention is possible. Not just by using the right combinations (see chapter 3) but also by sowing at the right time.

If you sow before the end of April or after the 15th of June and preferably in a windy spot, you will run far less chance of a carrot fly problem than the gardener who sows in May.

Once the carrots are big enough to be harvested, they should not be left in the ground for too long, as they offer an open invitation to the fearsome flies. Therefore you should lift the carrots as soon as possible and store them in a cool, dark place in moist peat or sand.

Shallots

Sowing shallots between the rows of carrots is another good way of avoiding carrot fly damage. The acids produced in the roots of shallots are lethal to the maggots of the fly.

PART TWO

How to Keep Pests
Out of the
House

Chapter One

Your house

Avoiding pests and plant diseases is always much better than having to fight them. Unfortunately people tend to overlook this — inside as well as in the garden.

Indoor plants often suffer unnecessarily, either because they are in the wrong place or because they are kept too wet or too dry.

Mistreatment will cause the plants to decline, resulting in an immediate attack by insects or fungi on the weakened tissues. As we have stressed before, pests will not bother strong healthy plants, because the cell walls of these plants are too hard for them to penetrate.

If you want to prevent diseases of plants in and around the house you will have to pay special attention to the following points:

Place the plants properly, shadow plants should be away from strong light, and even plants that like sunshine may not be able to withstand the strong rays just behind the glass.

Take care that the pot soil is kept moist, but for most plants not too wet. Also pay attention to humidity of the air. Unless you keep only cacti, this should be fairly high.

A good book on indoor plants, and sound advice when you buy a new plant, can save you a lot of trouble.

As far as flying and crawling insects are concerned, try to keep them out of the house, rather then having to destroy them indoors. This is always far more troublesome, especially if you want to avoid using chemicals.

Chapter Two

Aphids again

Biologists invariably wax lyrical at the mention of aphids, because these delicate little insects have such exceptional features.

One of these is that the females can reproduce for generations without the help of any male; matriarchy in its purest form. Ants keep aphids in the same way we keep cows but not for milk, for their sticky, sweet secretion.

If a plant becomes overcrowded by aphids, they react by producing a winged generation, which takes off and is carried on the air while searching for a likely plant. This excellent solution to their population problem has given the aphids the reputation of coming 'on the wind'.

But even if the aphid is a biological wonder, on your indoor plants it spells disaster. Not only do the plants suffer from the speed with which mother aphid and her innumerable offspring suck the sap from their leaves, they also get their air inlets blocked by the aphids fouling up the pores in the leaves with their syrupy excreta, which in addition forms an ideal medium for mould growth. It will be clear that it is essential to do something about these interesting animals with their explosive growth power.

This is not very difficult as the aphid is such a vulnerable creature.

Nettle Manure

An effective remedy against aphids is nettle manure (bunch of nettles in a bucket of water, cover and leave for five days, then sieve). But the ingredient for this spray is only available during the summer, which is fine for use on outdoor vegetables, as they grow only during that period anyway, but indoor plants may require some aphid treatment during winter and early spring as well.

Soap and Methylated Spirits

An old reliable method is the use of soap and methylated spirits which must be used sparingly, because meths is an alcohol which dissolves the green pigment in the leaf. More-over, the soap smears the leaves, so it is advisable to rinse them the following day.

Make the soap meths by mixing together:
2 pints of water
1 teaspoonful of soft soap
1 tablespoonful of methylated spirits.

Then spray on the infested plant.

Onion Extract

The onion can also be of assistance against the aphids during the winter. Use a pound of onions and one pint of water. Shred the onions finely, and let them simmer in the water in a closed container for one hour. Filter or sieve, and your spray is ready.

Nettle Extract

Ordinary nettle extract is also effective against aphids but in contrast to the nettle manure, which also works as a preventa-tive through the release of silicic acid, the extract only kills the aphids.

For this you take two pounds of freshly-cut nettles and leave them for 24 hours in a gallon of cold water. This mixture has to be applied generously in order to kill all the insects.

Repeat the treatment after a week to catch any persistant offenders.

Smoke

A very neat and quick way to free a plant from aphids without spraying is to put the plant in a pan or bucket that can be closed off tightly. Blow some cigarette smoke through a small opening and quickly seal the container. Leave this for an hour and then check the aphids for any signs of life. More likely than not they will all be stone dead. Just to be sure you can repeat the treatment about five days later. Remember one little aphid can produce a whole generation of hungry mouths in no time.

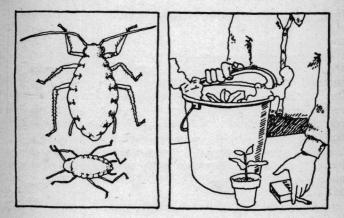

Artemisia Absinthium

An extract made from wormwood, *Artemisia absinthium*, as described in the chapter on aphids in the garden, will obviously be effective on indoor plants as well.

Chapter Three

The cockroach

In the past cockroaches were virtually unknown in this country. Being tropical animals they cannot stand the cold, and the ones we find here originally came from warmer countries. They stowed away in ships' cargoes and most of them soon died of cold. Only in places such as hotels and bakeries where it was warm all the year round, did they survive and multiply. Recently they seem to have discovered centrally-heated flats and houses. More and more flat-dwellers complain about cockroaches making their homes in the boilerhouse and crawling along the heating pipes.

These flat, brown insects, which can grow quite large, are fond of cereals and legumes. As with mice, their gluttony is not so objectionable as the way they pollute everything around them while they are eating.

The simplest method to keep them away is to lock everything edible out of their reach. But there are other means of getting rid of them.

Pickled Herring

Put some pickled herring, which they dislike intensely, either where they enter the room or under the heating pipes and radiators. They will disappear in no time.

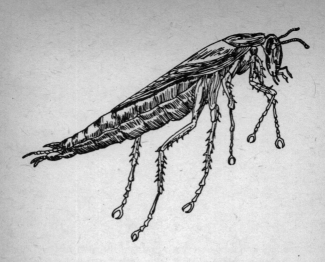

Borax

Make a mixture of Borax (poisonous to children!) and soft brown sugar, and put it near their breeding places.

Cucumber Peel

Some people believe that it also helps to put fresh cucumber peel in any chink or crack in the wall. If it doesn't do any good, it certainly cannot do any harm.

Chapter Four

Lice

Lice are Staging a Comeback

The common head louse, *Pediculus humanus,* is gradually making a reappearance. Every now and again you hear about outbreaks of lice in schools. With people generally wearing their hair longer again, it is much easier for the louse to spread. It likes a change of scenery at times and will try to find another host. Like the flea, the louse will stay on the same species; once used to human blood it will turn its nose up at cats and dogs, and vice-versa. Consequently humans have nothing to fear from lice on cats and dogs.

Be careful when trying to get rid of lice to also kill any young which hatch from the eggs left after the first treatment. Fortunately the louse takes more care with its eggs than the flea, which just lays them anywhere. The louse glues its eggs into the hair of its host, so we should know exactly where to look for them.

Vinegar

A simple method which is very effective, although it smells rather unpleasant, is to wash the hair with lukewarm vinegar. No louse can stand up to this. The vinegar has to be left in the hair for at least half an hour before it is rinsed out. This can be a problem with cats and dogs because they hate having vinegar in their coats. Moreover, they can catch cold, so it is advisable to treat them in, say, a well heated bathroom.

Comb

Once you have cleaned the lice from the hair of your child or pet, it is a good idea to use an old-fashioned fine-toothed comb once a week to catch any lice that might have hatched since the treatment.

Potato Water

A very old-fashioned method to prevent outbreaks of lice on animals is to wash their coats regularly with the water in which you have cooked the potatoes.

Paraffin

Human heads can also be de-loused with ordinary paraffin. Massage it well into the scalp, wrap a towel round the hair and let the paraffin soak in for at least half an hour. Next, wash the hair carefully with shampoo, so that no trace of paraffin remains.

This method is unsuitable for dogs and cats because they may lick their coats, and become ill from the paraffin.

Chapter Five

Ants

The little insect you may see toiling in your kitchen or
pantry is usually a worker ant. The ant community has a large
proportion of working-class members. Their task is to build
the nest and maintain it, hunt for prey, feed the larvae, 'milk'
the aphids and when necessary defend the nest as well. In
short, they take care of everything except reproduction. That
is the privilege of another group of ants: the males and the
queens.

These are winged, and mate during the so-called nuptual
flight. After the mating the males die but for the queen life
is just beginning. Coming down to earth she bites off her now
redundant wings with her powerful jaws and begins to work.
She digs a little hole in the ground in which she lays about
ten eggs. She looks after these eggs herself and feeds the
larvae which hatch from them. But as soon as they develop
into adult workers they take over the mothering task, and the
queen's only remaining job is to lay the eggs. This is the
reason why you seldom see a queen; she lives deep inside the
nest. The males, who have such a short and pleasurable life,
will hardly ever be troublesome in the house. However, the
worker ants will come indoors, in search of food for the
voracious larvae back in the nest. As they do more good than
harm by preying on other insects, you should leave them
alone as far as possible. If you want to prevent them coming
in, take care that all sweet foods, like sugar, jam, and honey
are put away and never spilt on the floor. Then there will be
nothing to attract the ants.

Try to be tolerant, even if you dislike the insects in your home, and keep them at a distance rather than killing them; they have a too important role to play in nature to justify such an action.

Put plants they dislike around the house, (on the balcony, the window-sill or in your garden. Here is a list of plants that will serve for this purpose.)

Lavender *(Lavendula Officinalis)*

This plant, with its fragrant blue flowers is very attractive, both in summer and winter. However, it will only thrive on calcareous soil, so if your soil is rather acid it is a good idea to dig in some crushed eggshells around the roots from time to time. These shells contain a lot of calcium, which is slowly taken up by the soil, thus creating the right conditions for the lavender.

Marigolds *(Calendula Officinalis)*

These annuals will also keep the ants at a distance, but only when they flower. The single variety is more effective than the double. Anyone scared of ants should make some early sowings under glass, to bring the flowers out as soon as possible.

African Marigolds *(Tagetes)*

As you will have noticed, African marigolds are good for everything. These little annuals fight eelworms in the soil, flies in the house, and ants both indoors and outdoors.

In contrast to the English marigold, they are effective against ants even before they flower. Of the many varieties of *Tagetes* that exist, *Tagetes patula*, sometimes called the French marigold, disposes of insects most effectively. A few rows in strategic places will form a good defence line against ants the whole summer through.

If you feel you must destroy the ants after all, here are a few ways in which to kill them, but once again only use these methods if the ants are really doing too much damage.

Camphor

Camphor is a crystalline substance derived from the sap of the camphor tree (*Cinnamamon camphora*). It has a very strong smell and is lethal to many insects. A teaspoonful of camphor in the nest may hopefully see the ants off, but will probably kill them.

Naphthalene

People often confuse naphthalene with camphor. This is a pity, as camphor is a natural product and comes from living trees, whereas naphthalene is made from coaltar. Naphthalene also has a very pungent smell and is a well-known insect repellent. A few spoonfuls around the nest will be enough.

Yeast

Strange as it may seem, ordinary baking yeast is lethal to ants. A saucerful of yeast and syrup, mixed together in equal amounts and diluted with water if necessary, will be briefly appreciated by the ants. Their sweet tooth will soon prove fatal.

Borax

Borax or sodium perborate (obtainable from chemists) is also lethal to ants. Because an ant will have more sense than to eat borax on its own, you will have to mix it with icing sugar. A lot of ant powders sold commercially have borax as their active ingredient, but you had better not use these, as they often contain other more harmful chemicals. Be very careful with the borax mixture when there are children about, as it is dangerous to them as well.

Some more Defence Plants

Some other plants that ants will avoid are:
tansy (*Tanacetum vulgare*), stinging nettle (*Urtica dioica*), spearmint (*Mentha crispa*), and chives (*Allium schoenoprasum*).

Chapter Six

The moth

The Moth is an Expert in Camouflage

The ordinary clothes moth is a master of camouflage. It
prefers dusk and twilight for flying in search of a suitable spot
to lay its eggs. The eggs themselves are so small that only the
sharpest eye can spot them. And the little caterpillars which
hatch from these eggs hide cleverly in cocoons made of the
very material to which they are attached. Consequently they
are perfectly disguised; and this makes it virtually impossible
to detect whether there are any moths in the wardrobe or not.
Only when they have fully developed into adults and flown
away, will their former occupation become evident from the
holes they leave in woollen and silken materials. By that
time though it is too late to take any action; the damage
has been done. Thus prevention here is highly important,
particularly the correct sorting of your winter clothes.

Air and Sunlight

First, you must air all woollen and silken garments and hang
them in the sun. The caterpillars of the moth loathe the sun
and will take their leave. After the sun has done her work, the
clothes should be brushed down and stored out of the moth's
reach.

Mothproof Bags

The old reliable mothproof bag will oblige here. They are mostly made of plastic nowadays and the fastidious moths don't like this material. So once the clothes have been out in the sun and are thus caterpillar-free, they can be stored in these bags without any risk of moth damage.

Mothproof Chests

Older housewives may still own one of these old-fashioned moth chests. Ideally they should be made of camphorwood, which is insect-repellent by nature, but more often they are formed from ordinary deal or good solid oak. In either case they must be properly sealed, because the tiny moth can squeeze through the smallest crack.

To be completely safe it is best to make the contents unattractive to the moth and there are several ways of doing this.

Camphor and Naphthalene

Every good chemist should still sell these two products. Check, though, that you are not sold a substitute, which may contain one of our infamous polluting chemicals. Insist on camphor (which comes from a tree) or napthalene (made of coaltar).

Place some of this on the bottom of the chest before putting in your clean clothes. The smell will deter the moth from laying its eggs on them.

Turps

Real turpentine is a natural product, made from pine resin. Do not confuse it with white spirit, which is used as a turpentine substitute and is made from petroleum.

Put a little in the last rinse when you wash your woollen jerseys, scarves etc. To the moth this trace of turpentine will make them unsuitable as a home for her offspring. Of course it can do the clothes no harm if you wrap them in plastic as well, as an extra safety precaution.

Saucers with Turpentine

Another method to keep moths away, is to place a saucerful
of turps in the bottom of the wardrobe where you keep your
woollies. This is very handy when you go away for a couple
of days and do not want to wrap everything up. It is simple,
effective and safe to the environment.

Tineola nisselliella

Newspapers

Printing ink is another repellent, and wrapping your clothes
in newspaper will effectively safeguard them from moth
damage. Be careful with light colours though; the ink can
come off quite easily.

Leaves of Certain Plants

The moth has a very acute sense of smell, and will avoid
places that smell of the dried leaves of walnut trees, wormwood, lavender, mint, or rosemary.

Paraffin

You can also spray with paraffin, but use this method
only for dark winter clothes and then sparingly, as paraffin
stains easily.

Chapter Seven

Gnats

The Male Gnat Does Not Bite

Gnats (*Culididae*) are not as aggressive as they are often made out to be. Many varieties, such as the little gnats you see buzzing about in great swarms in summer, do not bite or suck blood at all. Moreover, of the gnat families that do bite, only the females are dangerous. They need the blood for the formation of their eggs. The males of the same species live peacefully on plant saps and will not harm anybody.

As the larvae need water for their proper development, the eggs are laid on the still water surface of a pond or puddle. The fact that these larvae form an important link in the food-chain of both fish and frogs means that you should never use poisonous sprays to eradicate gnats. If a gnat survives an attack, her eggs, and hence the larvae that hatch from these, will contain the poisonous insecticide as well. Consequently, your action will eventually harm the predators of these larvae: fish and frogs. So anyone interested in the preservation of the environment does well to abandon these sprays.

Fortunately there are many ways in which we can defend ourselves against the unfriendly females.

Draught

The gnat is a weak and vulnerable insect, and very susceptible to draughts. Consequently the most effective way to get them out of your house, is to open the windows and doors for a while, so they are simply blown away. What could be more straightforward?

Screens

In the past, people used to have fine gauze screens in their windowframes, which allowed fresh air into the room, while keeping gnats and flies outside. It is time these screens made a comeback.

Camphor

Camphor, produced by a tropical tree, is an insect repellent which does not upset the natural equilibrium. You can quickly purge your bedroom of flies by putting some camphor crystals on the warm footplate of an iron and passing this along the walls. The gnats will die quickly and you will have an undisturbed sleep.

Castor-Oil Plant

The Ricinis- or Common Castor-oil plant (*Ricinis communis*) is an annual that needs a lot of space. It can grow up to six feet high and has enormous leaves. If you plant one of these near your window or door, you can be sure that no gnat will dare to enter. The smell of the castor-oil plant is very unpleasant to most insects, and even a few leaves in a vase will banish flies from the house.

However, families with young children are ill-advised to use this plant, as the seeds in particular are extremely poisonous. One single seed can prove lethal to a child.

Lavender Oil

Lastly some sweet smelling advice:
before you go to sleep, sprinkle your pillow with a few drops of lavender oil. It will keep the gnats away and give you sweet dreams.

Lemon Oil

This has the same effect as the lavender oil, but some people object to the strong lemon scent.

Chapter Eight

Mice

You will find the common grey mouse (*Mus musculus*) in nearly all older houses. There she makes use of any cracks and holes to build her nest. She will produce at least four litters a year, each consisting of about six little mice. These grow up quickly and will start to build their own nest within three months. So a young mouse couple moving into your house can certainly cause you problems. To provide for their family they will nibble at anything: paper, old shoes, clothes, cereals, legumes, meat and cheese are all appreciated. Moreover, they are not very particular about their eating habits and make a mess wherever they go. They don't feel any shame about their rapid digestion! So if you would prefer books without holes and unlittered kitchens, you will have to take some action.

A Cat

The simplest and most effective way to get rid of mice is to get a cat, not a pedigree but a common domestic moggie. Even a young kitten will be a good mouse deterrent. The hateful smell of cat will drive the mice away. If you are overrun with mice it is better to keep the cat on bread and milk for a while, and let it find its own supply of meat. However in the ordinary town house which is only occasionally visited by mice, you will have to feed your cat properly. A meat-free diet is really only suitable for farmcats, which have a much

larger hunting ground.

Common Spearmint

Mice absolutely hate the smell of Spearmint (*Mentha spicata*). Put some leaves of this plant in the mousehole (in winter the dried leaves will be effective) and the mice will soon run to escape the penetrating odour.

Peppermint Oil

The peppermint plant (*Mentha piperita*) is closely related to the spearmint. It's strongly scented oil is obtainable from any good chemist's. A few drops on cottonwool will have the same effect as the spearmint leaves.

Sunflowers

The sunflower not only gives us beautiful flowers and valuable seeds, which contain a high quality oil (rich in unsaturated fatty acids), but can also be used as a pesticide against mice. Harvest the flower heads before the seeds are completely ripe, and so still firmly embedded in the receptacle. After they have been dried in a suitable spot, cut each head into four and put the quarters near a mousehole. The unsuspecting mice will nibble gratefully at the seeds, but will also take in small amounts of the receptacle, which is poisonous. It is rather a cruel method, but at least it does no harm to any other creatures.

Cement Mixed with Glass

An old and well-tried method is to close the entrance of the hole with a mixture of cement and bits of glass. It is the glass that does the trick, because the mice will not have the courage to gnaw their way through it, and will have to look for another spot with easier access.

Chamomile and Oleander

Among the smells that mice detest are also those of the dried and rubbed leaves of wild chamomile (*Matricaria chamomilla*) and oleander (*Nerium oleander*).

Rags Drenched in Tar or Creoline

As with rats we can try to frighten the mice away by using rags drenched in tar or creoline, pushed into the holes.

Shrews

While talking of mice, we should also mention the shrew. This little animal, which usually lives in the country, does no harm at all, as it feeds on insects and never touches any other food. So if you ever come across one, you had better leave it be.

Chapter Nine

Woodlice

However strange it may sound, the small greyish woodlice (*family Isopoda*) is related to the crab, the shrimp and the lobster. You could say that it is a miniscule shrimp which has adapted to living on the land. It still shows preference for moist surroundings though. The woodlouse has seven pairs of sturdy little legs, with which it can move very fast, and when threatened it quickly rolls up into a ball.

The woodlouse lives on decaying vegetable matter, and is therefore the most harmless animal imaginable. It does a useful job in cleaning up all the vegetable rubbish it comes across, so there is no reason whatsoever to destroy it.

Why then a chapter devoted to this insect? Because there are housewives that nearly faint at the sight of a poor innocent woodlouse, and immediately jump for the spray can to kill it. In this way they do more harm than good, because all these chemical insecticides add to the pollution of the environment.

If you really want to get rid of the woodlice, then cut a potato in half, hollow it out and make a small opening in the sides, so that you form an 'igloo'. Place this in a spot where the woodlice usually are. They will like the little house and creep inside. After a day or two you throw the whole thing in the garden. Just repeat this till they have all gone. But again, why not leave them be?

Chapter Ten

The spider

Spiders (family *Arachnida*) are amazing insect hunters. Most varieties snare their prey in cunningly devised webs, but some hunt in the open, and suddenly leap on top of their unsuspecting victim. In this way they do a very useful job by keeping the numbers of (mainly flying) insects down. Therefore the Scots saying that it is unlucky to kill a spider is very sensible.

Unfortunately, many people find spiders unattractive, sometimes even frightening, while housewives resent them weaving their sticky webs in nooks and crannies. So they are inclined to destroy them, often with some dangerous insecticide. Two mistakes in one go!

The best method is to catch the spider in a cloth or on a broom and take it outside, though a clever spider will soon find his way back in. To prevent this, you will have to make the corner where it has its web as unattractive as possible.

Sulphur

That can be done quite easily by spraying some sulphur powder in its favourite spots. This will keep the spider away, without killing it.

Chapter Eleven

Red spider mites

If the undersides of your houseplants are covered with a white veil which slowly turns yellow, this is due more often than not to an infection by the red spider mite of the family *Tetranychidae*.

This minute spider makes its web on the underside of the leaf, and breeds there. The entire family feeds on the sap of the plant, which duly suffers from this attack.

Unfortunately, it is very difficult to control this tiny insect once it has established itself. Therefore it is best to prevent it by keeping the air humid and sponging the leaves regularly with water: a red mite infestation is nearly always caused by air that is too dry, or an environment which is too warm.

Nettle Manure

Once the mite has settled on your plants, it will take a lot of perseverance on your part to get it off again. You will have to treat the plants several times, and even then you may not succeed. However, your best bet in this struggle is to use nettle manure as mentioned before. Apply it generously, and repeat the treatment once a week to increase your chances of success.

Mixture of Soap and White Spirit

If nettles are out of season, you can also try a mixture of
soap and white spirit:

two pints of water, a teaspoon of soft soap, and a tablespoon of white spirit. Rinse the leaves with either rainwater
or boiled water an hour after the application.

Derris and Pyrethrum Powders

A mixture of derris and pyrethrum powder will also help, but
again it needs to be applied several times. A really good
garden shop should be able to supply it.

Chapter Twelve

The fly

Why the Fly is Important for the Preservation of the Environment

Nature uses her resources far more economically than so-called *Homo sapiens;* she sees to it that any waste is cleaned up and quickly converted into useful material. The common housefly (*Musca domestica*) for example is an expert in this 'recycling': she lays her eggs on preferably decaying plant or animal remains, which serve as food for the developing larvae and are returned to the soil in the form of their droppings. They then form a foodsupply for plants and bacteria and so the cycle is complete.

Thus the fly is indispensable to the economy of nature. Unfortunately, it has other less desirable attributes as well. It can be very dangerous and many people justifiably hate flies, because they act as vectors for innumerable diseases, conveying germs from one host to another. However you should never fight flies with chemical insecticides, because you indirectly harm the environment by doing that: birds that live on flies, and the soil that harbours the larvae, will eventually be poisoned by the chemical.

Prussian Blue

Our grandmothers knew and exploited this fact: flies hate the colour blue, and will never settle on it. Why, nobody knows, but kitchens painted blue will stay free of flies. This method is even more effective if you mix some finely ground alum into the paint. It is the simplest and friendliest method to keep flies at a distance and particularly recommended for farmers, who want to keep their stables and livestock sheds free of flies.

Screens and Beadcurtains

Of course we can also keep the flies outside by fitting fine gauzed screens in our window frames or by hanging decorative beadcurtain in the kitchen doorway. The slight clattering of such a curtain is just enough to discourage flies and other flying insects, like wasps and gnats.

African Marigolds

The African or French Marigold (*Tagetes patula*) is an amazing plant. It appears in so many different chapters of this book because it will assist you in fighting just as many different pests. Other single varieties of *Tagetes* also have this insect repellent effect, though usually to a lesser extent than *Tagetes patula*.

A few of these plants on the windowsill, or in a windobox outside, will not only look very bright and colourful, but will also keep the flies away. Water them regularly and give them some compost now and again.

Lemon scented Pelargonium

This special variety of *Pelargonium* is an ideal solution to your fly problem. It has deeply lobed leaves and spreads a very pleasant lemon scent, which only flies could find disagreeable. Provided it gets some sunlight, either in the morning or evening, regular water, and some compost now and again, it will grow to a sizeable and very decorative plant.

Southernwood *(Artemisia Arboratum)*

This member of the family *Compositae* is a close relative of Wormwood. It is an old-fashioned plant, often found in more established gardens. It has small glands which contain lemon scented essential oil, an ideal fly repellent. A few branches in a vase will keep your rooms free of flies.

Elder *(Sambucus Nigra)*

Old farmhouses often have elder trees near the kitchen or the out-buildings, because farmers are aware of their effectiveness against flies. Of course you cannot grow an elder tree in your garden overnight, but some elderflowers in a vase will do the job just as well, though you will have to renew them every three days or so.

Walnut Trees *(Juglans Regia)*

Here is some good advice for people with a large garden, especially on clay soil:

plant a walnut tree near your kitchen door. One tree will be sufficient because flies cannot abide the peculiar nutty scent that it spreads. Again a few branches indoors will help.

Onion Water

Flies often annoy housewives by depositing their excreta on mirrors and windows. A simple way to forestall them is to wash your windows with onion water:

Cut up one pound of onions very finely and pour two pints of boiling water over them. Let this mixture cool down, then strain it. Rinse your chamois leather in this extract while you are doing the windows. Flies will carefully avoid any glass treated in this way.

Nettles and Tansy

Two other plants that flies dislike intensely are nettles (*Urtica dioica*) and tansy (*Tanacetum vulgare*). In the past tansy was often grown around farmbuildings to keep flies off the live-stock. Nettles are equally effective (though not very attractive).

Slices of Lemon

On a warm summer evening it is a good idea to put some freshly cut slices of lemon on the windowsill. These will form a scent barrier which flies will not cross.

Wormwood Spray

A fly spray we can use safely — that is without harming the environment is one of wormwood (*Artemisia absinthium*). Pick some fresh wormwood leaves, cover with water and bring to the boil. After this has cooled down, sieve it and dilute in a proportion of one to four. Stir the spray well before use: ten minutes, alternately clockwise and anticlockwise.

Chapter Thirteen

The flea

The Flea is a Slovenly Insect

The female flea is the most careless mother you can imagine!
She lays her little white eggs at random, without attaching
them to anything. In most cases they just fall to the floor
where, because they are minute, they are scarcely visible and
all but impossible to sweep up. Consequently, the floors of
houses where there are pets plagued by fleas will be littered
with tiny eggs, which in due course will produce the larvae.
These feed on whatever animal material they can find: a
woollen carpet is ideal. Once they have grown fat, the larvae
spin a cocoon from which eventually a tiny flea will emerge.
If necessary these young fleas can remain inactive for weeks,
even months, in order to conserve energy; and only when they
suspect prey near at hand will they start to jump.

This patient inactivity of all young fleas accounts for the
widespread flea epidemics which occur during the warm
summer months, particularly in houses where there are dogs
and cats. During holidays the eggs develop into fleas which
wait quietly for the residents to return and then make their
attack *en masse*.

If they are dog or cat fleas your pets will suffer accordingly;
these will usually avoid humans, because the different species
tend to stick to their favourite host. Occasionally, however,
they will take a bite at any warmblooded creature, animal or

human to whom their erratic leaps have brought them. The same holds true for the human flea; it will not bother your pets, though you will naturally do all you can to be rid of this nasty insect.

The best thing to do, of course, is to avoid this situation, in other words, vacuum, sweep, and if possible mop the floor with a mixture of soft soap and meths, before you go away on your holidays. Then you will nip the problem in the bud. The dog's basket and the cat's favourite cushion should also be cleaned thoroughly, as they are the flea's breeding-ground.

Comb

One of the simplest and most sensible ways to fight fleas is to catch them with a fine-toothed comb. Not a very appetising occupation, but a very effective one as fleas are becoming more and more resistant to insecticides. However, as pointed out in the introduction, combing alone is not enough; you still have to get rid of the eggs on the floor, which create far greater problems than the fleas themselves. Beat and vacuum carpets and mats as thoroughly as you can and as often as possible.

Derris Powder

Derris powder, which contains rotenone as its poisonous ingredient, is a natural insecticide, produced by tropical *papillionaceae*. The roots particularly contain a high percentage of this rotenone, which kills insects on contact.

The powder has to be rubbed into the coat of your dog or cat. Although derris powder is not actually poisonous to mammals, it is not very good for them either and therefore it is better to rub it into places that cannot easily be licked; put it on the neck and behind the ears, rather than on the back or the belly.

This derris treatment has to be repeated every ten days which is the time it takes for the new fleas to emerge from their cocoons until they have all disappeared.

Pyrethrum

This is a composite flower, the effect of which has been long appreciated. Great grandmother dried its flowers and sewed them into small muslin bags for use against all sorts of insects.

If you are beleaguered by fleas, it is advisable to put a few of these little bags in beds, and on the cushions where your pets usually sleep.

Pyrethrum
(Chrysanthemum)
coccineum)

Wild Chamomile

Wild Chamomile *(Matricaria Chamomilla)*

This composite has the same effect as pyrethrum, but to a lesser extent. Therefore you will have to use it more generously. On the other hand, it has a very strong scent and regular contact with it will make the skin of your dog or cat less attractive to the blood-thirsty fleas. One advantage of chamomile is that it is easily obtainable. A good chemist should be able to supply the dried flowers, but you can also gather the fresh ones yourself during the summer. They grow in abundance on verges and cultivated land. While doing this, you have to be careful not to confuse the wild chamomile with its scentless cousin *Matricaria inodora*. The difference between these two is found in the receptacles, which in the wild chamomile are coneshaped and full of tiny yellow flowers, sometimes surrounded by a circle of white ones. If you are not quite sure, consult a book.

Bog Myrtle or Sweet Gale

Bog myrtle (*Myrica gale*) is a fragrant shrub with narrow grey-green leaves. Its favourite types of habitat are fens and wet heathlands. In spring it produces upright catkins, which develop into small fruits during the summer; it is these fruits that you need. Fleas cannot stand their smell, and sewn into a bag, together with some dried leaves, they will be very effective in keeping your pet's basket free of fleas.

You will have to take the trouble to find the shrub yourself though, because its fruits cannot be bought.

Ferns

An ancient method of getting rid of fleas is to stuff the dog's or cat's cushion with dried ferns. Ferns are said to frighten fleas off. It is worth a try.

Soapy Water

All the methods outlined so far dealt with fleas in the coat of your pets and will not solve the problem of those fleas that still shelter in your carpet because they have not yet found a suitable host. Apart from the vacuum cleaner, there is another, more romantic way of coping with this category.

Take a soup plate and fill it with soapy water, pour a layer of oil on top and put a candle in the middle. Place this whole contraption on the floor, and light the candle when the room itself is in complete darkness.

If this works properly (and it usually does) the plate will

be full of drowning fleas within a few hours: the fleas are attracted to the light, jump in the plate, stick to the oil, and drown in the soapy water.

You will have to use this 'flea trap' several times in order to catch any newly hatched fleas.

Wormwood

You can also bath your pets in an extract of wormwood (see chapter on aphids). It is harmless to your animals but lethal to fleas.

Mopping

In rooms frequented by your pets it will be necessary to mop the floor regularly with water which has had creoline or alum added to it. This kills both the eggs and the larvae.

Carpets and mats can be sprayed with one of these solutions. It is wise to test the carpet material first, in case of any ill effects which the treatment may have on a large scale.

Chapter Fourteen

The wasp

The Wasp Builds a Paper Nest

Although the wasp (*vespula vulgaris*) is an insect which many
people fear, she will only sting when she feels threatened.
Therefore the best way to avoid being stung is to sit very still
and not to move until she flies away again. (In most cases it
will indeed be a 'she', one of the female workers, as the males
who never sting, have a much shorter life span).

Wasps, like ants, live in a well-organised community, though
the wasps' nest is quite a bit smaller than that of the ants:
there are usually two to three hundred wasps per nest.

In spring, when the females come out of their winter
sanctuary, they start building the nest in which they will lay
their eggs. Under their mothers' watchful eyes these first eggs
develop into larvae and subsequently into adult wasps which,
destined to be workers from the beginning, start work as
soon as they have hatched out. First of all they enlarge the
nest. The raw materials used for this are bits of paper and
wood splinters, which the wasps chew patiently until they
form a pulpy mass. Out of this they build new hexagonal cells
onto the nest, ready for the queen to deposit her eggs. The
resulting larvae are fed on insects, and once mature they join
the army of workers. In this way the nest keeps expanding
throughout the summer.

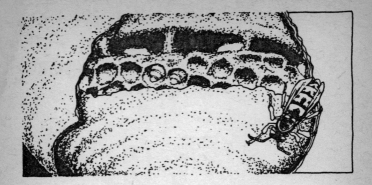

Although the larvae are fed on animal matter, wasps, like ants, are very fond of sweet substances. The smell of jam or apple sauce simmering on the stove, will attract whole swarms into the kitchen.

Full protection against wasps is hard to achieve, though the fine gauzed screens mentioned before will be quite effective.

Ammonia

Another method to keep the wasps at a distance, is to put out some saucers filled with a mixture of water and ammonia, the smell of which wasps cannot abide.

Chapter Fifteen

Unpleasant smells

Over the course of centuries our sense of smell seems to have become more and more acute, until we have reached the point where we are disturbed by the smell of cooking or tobacco smoking. In most cases people resort to so-called 'air fresheners' in spray cans, which drown one smell with another, and only add to the problem of air pollution. Proper ventilation will be far more effective, and cheaper at that. Fresh air is still free, though admittedly hard to come by in some places.

Vinegar and Lemon Juice

It will often help if you put a saucerful of vinegar or lemon juice next to the cooker when you are preparing a meal. This will keep the air fresh by absorbing the fumes. It goes without saying that you cannot use the same vinegar for weeks on end; you will have to replace it every so often.

Apples

Besides being tasty to eat, a fragrant ripe apple has other good qualities: it can be used to lessen the pungent smell of boiling fish. Just put a piece of apple in with the fish. (If you throw it out for the birds afterwards, they will certainly appreciate it!)

White Bread

A slice of white bread can remove many nasty smells. When you are making chips, for instance, a few pieces of bread in the chip pan will take away most of the fumes.

My grandmother always used to put a slice of dry white bread on top of the cauliflower, sprouts or cabbage she was boiling. If you want to use this method, you should not use too much water as the bread should stay fairly dry. The main thing with cabbage, however, is not to overcook it. Not only does that spoil the texture, it also brings out that awful, rank cabbage smell.

Cigarette Smoke

Cigarette smoke can undoubtedly spoil the atmosphere in a room. The best remedy of course, is to ask your family and guests not to smoke, but that is rather unsociable.

One way to alleviate the effects of smoking is to burn candles in the room. This will look very attractive at the same time.

A bowl of water, placed inconspicuously in a corner, will also help, by absorbing a lot of the smoke, especially if you add some vinegar to the water.

Pomander or Clove Orange

An ideal way to maintain a fresh smell in a cupboard or room, is to hang up a deliciously scented pomander or clove orange. Our grandmothers were truly skilled in making pomanders, decorating them with silk ribbons and special patterns of cloves. Our recipe is somewhat simpler: Select a ripe thin-skinned orange and stick it full of cloves, starting from the stalk-end and going round the orange until it is completely covered. Then wrap it neatly in tissue paper and put it in a dry, warm place. After a fortnight the orange will have shrunk considerably, but it will not have lost its fragrance. Also, the volatile oil of the cloves is a powerful antiseptic, which will permanently protect the pomander against mould attack. Finally tie a silky ribbon around it and your pomander is ready for use.

Herb and Flower Sachets

Another remedy against stale smells are sachets or sweet bags, filled with a variety of herbs or petals. Like clove oranges, these sachets make an original present for friends, particularly if you take the trouble to embroider them prettily and to select an attractive material like batiste, cambric or muslin.

Lavender sachets are well-known; they will keep their scent for at least a year.

Other herb sachets can be filled with a mixture of cloves, rosemary, and nutmeg. This will give a nice spicy scent. Herbs like southernwood, spearmint, savoury, and thyme are better used on their own.

If you are lucky enough to have some fragrant roses in your garden, you can fill your sachets with rose petals. Pick the flowers when they have fully opened. Do this preferably in the morning, after the dew has evaporated and before the roses have been exposed to the hot sun. Discarding the white parts, put the petals on some tissue paper, ensuring that they do not touch. Try to find an airy but shady spot, where they can dry quickly. If the weather cooperates, you can fill your sachets after a week's drying. Some people like to put a clove in with the rose petals, but this can overpower their smell.

Other possibilities are:
dried jasmine flowers (which also taste delicious blended with ordinary tea), honeysuckle, the leaves of the lemon-scented pelargonium and the leaves and flowers of bergamot (*Monarda*).

In short, there are endless opportunities for experimenting with the scents of different flowers and plants.

Index

Acid soil 7
African marigold 14, 16, 47, 60, 75
'Air fresheners' 85
Alum 32, 82
Allium schoenoprasum 61
Ammonia 84
Ants 52, 59–61
Aphids 16, 24–7, 52–4
Apple 85
Artemisia absinthium 26, 54, 77
A. arboratum 75
Artificial fertiliser 8, 12, 24
Asparagus 14
Astrological calendar 12
Azalea 7

Bad companions 17–18
Barberry 35
Bead curtain 75
Beans 14, 16, 17, 35
  broad 14, 16, 25
  runner 13, 14, 16
Beer 43
Beetroot 14
Berberis vulgaris 20 35
Bergamot 87
Birch 17, 26
Birds 37, 74
Blackfly 14, 25, *see* aphids
Bloodmeal 10
Bog myrtle 81
Bonemeal 10
Borage 16
Borax 56, 61
Box 18
Bread 86
Broad-beans 14, 16, 25

Brussels sprouts 86
Buxus sempervirens 18

Cabbage 13, 14, 16, 17, 18, 86
  collar 23
  root fly 22–3
  white butterfly 36

Calendula officinalis 60
Camomile 14
Camphor 29, 61, 63, 66
Carbide 29
Carrot 12, 14
  fly 48
Castor oil plant 66
Cats 31, 57, 58, 67
Caterpillars 36–7
Cauliflower 14, 17, 86
Celery 14
Cement with glass 69
Chalk 44
Chamomile, wild 69, 80
Chemical insecticide 74
Chervil 14, 27
Chives 16, 61
Cigarette *see* nicotine
  smoke 86
Cinnamamon camphora 61
Clove 86, 87
Cockroach 55–6
Comb 58, 79
Common rue 20
Companion plants 15
Compost
  making 9–12
  using 12
Conifers 21, 35
Creoline 82

Cress 14, 19
Cucumber peel 56

Darwin 33
Derris 36, 73, 79
Digitalis purpurea 21
Dill 14, 26
Dogs 20, 57, 58

Earthworm 9, 33
Earwig 30
Eelworm
  potato 14
  tomato 17
  tulip 17
Eggshell 44, 60
Elder 76
Equisetum 41
Euphorbia 31
E. lactea 29

Feathermeal 10
Fennel 17, 18
Ferns 81
Fertiliser 8, 12, 24
Fir seed 42
Fish 65, 85
Flea 78
  beetle 14, 19
  trap 81—2
Flower pot 30, 45
Flower sachet 87
French marigold 60, 75
Fritillaria imperialis 31
Frog 65
Fruit tree 12
Fungi 12, 40—1

Gladioli 17

Gnats 65, 75
Golden rod 14
Gooseberry 18
Grass mowings 10, 37
Guinea fowl 32

Heather 7
Herb sachet 87
Hemp 16
Herring, pickled 55
Honeysuckle 87
Horse-chestnut 34
  leaves 10
Horseradish 13, 16
Horsetails 41
House plants 51, 72
Housefly 74—7
Hyssop 13, 14, 26, 42
Hyssopus officinalis 42

Insect 12, 51

Jasmine 87
Juglans regia 77
Juniper 35

Kohlrabi 18

Ladybird 26
Lavender 8, 16, 26, 60, 64, 66,
  87
Lavendula officinalis 60
Leek 14
Lemon 77
  juice 85
  oil 66
Lettuce 14, 16, 17
Lice 57
  plant *see* aphids
Lime 8, 10

Maple leaves 10
Marigolds 13, 60
  African 14, 17, 60, 75
  French 60
Matricaria chamomilla 69, 80
M. indora 81
Mentha crispa 61
M. piperita 14
M. spicata 68
Methylated spirits 38–9, 53
Mice 67–9
Mildew 40
Mint 14, 64
Mole 28–9
Mole-cricket 45
Monarda 87
Moth 62–3
Mothproof bags 63
Myrica gale 81

Naphthalene 61, 63
Nasturtiums 47
Nettles 14, 17, 61, 77
  extract 25, 47, 53
  manure 24, 46, 53, 72
Nerium oleander 69
Newspaper 64
Nicotine 34
Nitrogen 10, 13, 16
Nutmeg 87

Oleander 69
Onion 14, 16, 17, 21, 26, 31
  extract 40–1, 53
  fly 14
  water 77
Orange, clove 86
Organic farming 9, 12, 13
  manure 8

Paraffin 58, 64
Parasite 7
Parsley 14, 17
Pears 35
Peas 13, 14, 16, 17
Pebbles 44
Pelargonium 75, 87
Penicillium 40
Peppermint oil 68
Petals, to dry 87
Phytophtera 16
Pine needles 17, *see* conifer
Plant combinations 13–18
Plant lice *see* aphids
Pomander 86
Potato 13, 14, 17, 70
  eelworm 14
  peelings 10
  water 58
Printing ink 64
Primula 22
Prussian blue 74
Pyrethrum 16, 36, 73, 80

Rabbit 21–2
Radish 14, 17, 19
Rat 31–2
Rhododendron 7
Rhubarb 43
  soap 25
Ricinis communis 66
Rose 7, 13, 16, 18, 26, 87
Rosemary 64, 87
Runner bean 13, 14 16
Rust 35

Sage 26, 42
Salvia officinalis 42
S. splendens 42

Sambucus nigra 76
Savory 27, 87
Scale insects 38–9
Seifert, Alwin 12
Shallots 48
Shrew 69
Silver fir 21, *see* conifer
Slug 42–4
  pellets 44
  poison 44
Smells 85
Smoke 54
  cigarette 86
Snails 42–4
Soap and methylated spirits
    38–9, 53
  and white spirit 73
Southernwood 75, 87
Sowing dates 12
Spearmint 61, 68, 87
Spider 71
Spinach 12, 16, 17
Spring onions 14
Spurge 31
  milk 29
Straw 44
Strawberry 13, 16, 18
Sunflower 16, 30, 68
Sulphur 41, 71
Swedes 14

Tagete 60, 75
T. erecta 47

T. patula 14, 47, 75
Tanacetum Vulgare 61, 77
Tansy 61, 77
Tar boards 19
  rags 32, 69
Thyme 26, 87, 42
Tomato 14, 17, 18
  eelworm 17
Treacle board 19
Tropaeolum 47
Tulip 17
  eelworm 17
Turpentine 63, 64

Urtica dioica 61, 77

Valerian 31, 34
Valeriana officinalis 31, 34
Vinegar 57, 85, 86
Vole 31

Walnut 64, 77
Wasp 83, 75
White spirit 73
Whitefly 46–7
Window screen 66, 75, 84
Wood ash 34
Wood-shavings 30, 44
Woodlice 70
Wormwood 27, 54, 64, 77, 82

Yeast 61
Yew 21